DINOSAUR MANDALA COLORING BOOK

JHL CODY PUBLISHING

Copyright © 2020 by JHL Cody Publishing
All rights reserved. This book or any portion thereof
may not be reproduced or used in any manner whatsoever
without the express written permission of the publisher
except for the use of brief quotations in a book review.

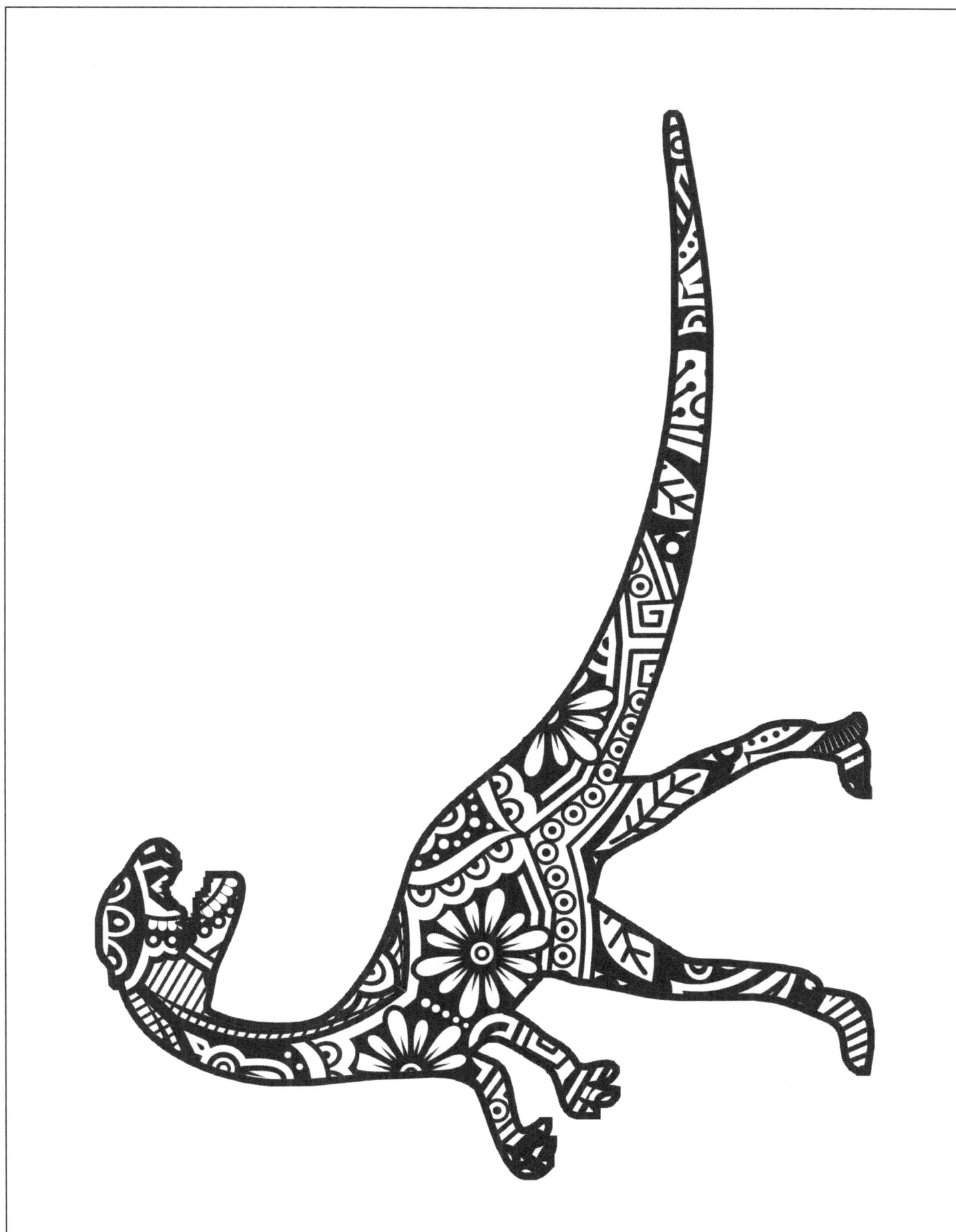

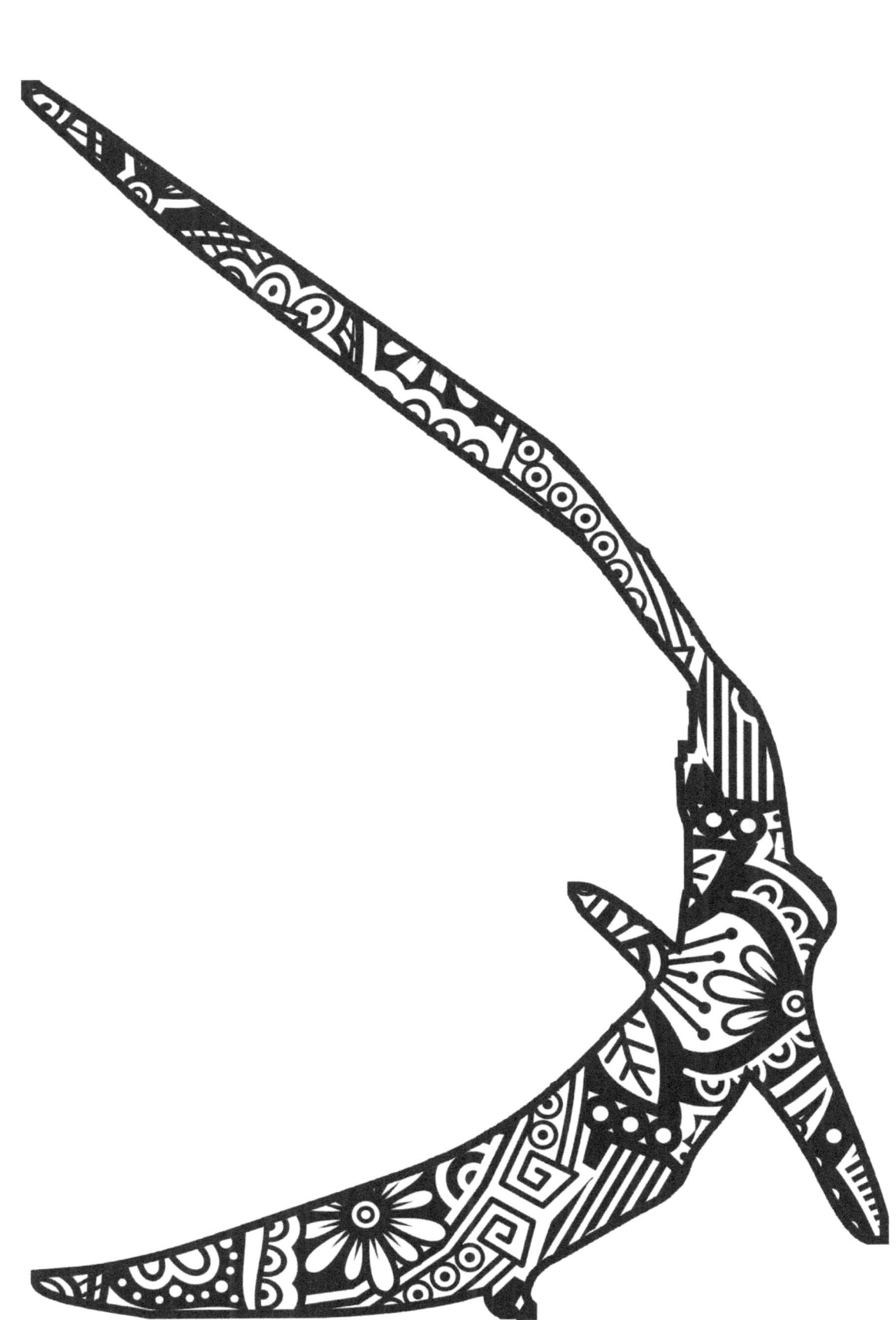